W9-DIN-322

Patterns in Poetry

Recognizing and Analyzing Poetic Form and Meter

Greg Roza

PowerMath™

The Rosen Publishing Group's
PowerKids Press™
New York

Published in 2005 by The Rosen Publishing Group, Inc.
29 East 21st Street, New York, NY 10010

Book Design: Haley Wilson

Photo Credits: Cover, background all pages © Creativ Collection; cover (Shakespeare), pp. 19, 28 © Bettmann/Corbis; p. 5 © Lester Lefkowitz/Corbis; p. 6 © Stapleton Collection/Corbis; p. 7 © Danny Lehman/Corbis; p. 9 © Stockbyte; p. 11 © Royalty Free/Corbis; p. 13 © Hulton-Deutsch Collection/Corbis; p. 20 © Peter Macdiarmid/Reuters/Corbis; p. 21 © Archivo Iconografico, S.A./Corbis; p. 23 © Michael Nicholson/Corbis; p. 24 © Nathan Benn/Corbis; p. 27 © The Bridgeman Art Library/Getty Images;p. 29 © Araldo de Luca/Corbis; p. 30 © EyeWire.

Library of Congress Cataloging-in-Publication Data

Roza, Greg.
Patterns in poetry : recognizing and analyzing poetic form and meter / Greg Roza.
p. cm. — (PowerMath)
Includes index.
ISBN 1-4042-2941-8 (lib. bdg.)
ISBN 1-4042-5146-4 (pbk.)
6-pack ISBN: 1-4042-5147-2
1. English language—Versification—Juvenile literature. 2. English language—Rhythm—Juvenile literature. 3. Literary form—Juvenile literature. 4. Poetics—Juvenile literature. 5. Poetry—Juvenile literature. I. Title. II. Series.
PE1505.R69 2005
821.009—dc22

2004003249

Manufactured in the United States of America

Contents

Patterns in Our World

A pattern is a repeating arrangement of elements or forms. A pattern is also a model for a way of making or doing something new that looks or sounds like the original. Patterns are **predictable**. This means we can guess what will come next because the pattern repeats. There are many kinds of patterns in mathematics. The numbers we use when we count are a good example of a pattern: each number is 1 more than the number that comes before it.

Patterns are everywhere in the natural world, from the seasons of the year, to the markings on butterfly wings, to the leaves on a tree. Many people think natural patterns display an orderly and even artistic beauty. People have copied nature by filling our world with man-made patterns. Many people think that repeated shapes—like those that make up the building shown in the picture on the opposite page—provide a sense of order and beauty for our man-made world that mirrors the order and beauty found in nature. The tasks that we do every day are patterns of repeated activities and actions. We wake up, we go to school or work, we come home, and we go to sleep. Then, the pattern begins again.

Patterns are found in all forms of human behavior. Dance is often based on patterns of repeated movements. Music features patterns of sounds that repeat to create a melody. Even the sports we play are based on patterns of actions, events, and directions.

Geometrical patterns are patterns made with shapes. This building includes patterns of squares and triangles.

The illustration shown here—a page from the *Book of Durrow*—is thought to have been created in Ireland about A.D. 675. This "carpet page" is covered by decoration, much like designs cover some rugs.

Patterns often play an important role in visual arts. Many paintings use repeating images of color and shape. Quilts are also considered works of art with their countless colorful patterns. The picture below shows cloth featuring the colorful stitched patterns of the Kuna Indians of Panama. This kind of art, called "mola," is often inspired by the natural world or the artists' dreams. You can easily see the pattern in this artwork.

Just as an artist may create a work with a repeating image, a writer may begin a work with a pattern in mind. Many poems are based on a **predetermined** pattern of words, **syllables**, lines, or **rhymes**. In fact, patterns in poetry can help to create emotions and themes. To understand the patterns we find in many poems, we need to understand the concepts of poetic form and poetic meter.

What Is Poetic Form?

Form is the shape and structure of an object. It is determined by the parts that make up an object. In geometry, shapes are forms. A triangle, for example, is a form that is determined by several features: it is a closed plane shape, it has 3 straight sides, and it has 3 angles. These features can be thought of as a model, or pattern, for all triangles.

Just as geometric shapes have predetermined forms, so do many poems. Instead of sides and corners, however, poetic form is determined by language and by how the poem looks on the page. A poet considers many word qualities when writing a poem, including rhyme, repetition, sound, length, and appearance. The lines that the words create also affect the form of the poem. Some poems have long lines, some have short lines, and some use a combination of the two. Other poems use space between the lines to help establish the form.

Good poets know how to use form to help convey the main theme of a poem. For instance, a poem with long, rhyming lines may help establish a songlike theme. On the other hand, a poem with short lines and words that can be spoken quickly may help establish a fast pace for an exciting theme.

Concrete poetry is a kind of poetry in which form is as important as word choice. In a concrete poem, the visual presentation—combined with the meanings of words—helps the reader understand the poem's theme. Look at the poem on the opposite page. The author used the form of a seagull as the pattern for the arrangement of the words.

of sea gull

beat beautiful wings

up fluid birds fluid up

down down

Both the words and the form of this concrete poem suggest the movement and shape of a seagull's wings.

Let's look at a specific kind of poem to see how form works. A cinquain (SING-kane) is a 5-line poem that always has exactly 22 syllables. Each line has a specific number of syllables. These features define the form of a cinquain. An example is shown in the box below.

The cinquain is designed to build **tension**, then release it quickly at the end of the poem. It starts with a short line, sometimes a noun that is the focus of the poem. The next 3 lines get longer and longer, building tension as the theme unfolds. The fourth line is the longest in the poem. However, the fifth line is short. This abrupt shift from increasingly long lines to a short line lets the poet release the tension and surprise the reader. You might compare this form to a balloon. As you blow air into a balloon, it gets bigger and bigger. The increasing amount of air puts tension on the balloon until POP! The balloon breaks, collapsing and quickly releasing the tension.

cinquain form		cinquain example
first line: 2 syllables	→	My breath
second line: 4 syllables	→	expands within
third line: 6 syllables	→	colorful rubber skin.
fourth line: 8 syllables	→	Careful, one breath too many and . . .
fifth line: 2 syllables	→	it popped!

The cinquain was modeled partly on a Japanese poetic form called the haiku (HI-koo). A haiku always has 3 lines and 17 syllables: 5 in the first line, 7 in the second line, and 5 in the third line. Traditionally, a haiku focuses on nature.

haiku example

**spring rivers stretch and
yawn, waking and escaping
winter's icy grasp**

cricket songs echo
softly between satin folds
of ripening wheat

Some poems are based on patterns of rhyming lines rather than—or in addition to—the number of syllables in each line. The pattern of a poem's rhymes is called its **rhyme scheme**. Poets use lowercase letters to map out a poem's rhyme scheme. An "a," for example, represents the sound at the end of the first line. The poet uses another "a" every time that sound is repeated at the end of following lines. New rhymes are represented by new letters.

A good example of a rhyming form is a limerick (LIH-muh-rik). A limerick always has 5 lines. The first, second, and fifth lines rhyme. The third and fourth lines rhyme, but they do not rhyme with the other lines. Often, but not always, limericks have a predetermined number of syllables in each line. The first, second, and fifth lines have 8 or 10 syllables. The third and fourth lines have 5 or 6 syllables. The rhyme scheme of a limerick is written in the box below.

Limericks are usually written about a silly or humorous theme. They are usually not very serious and are fun to write. Follow the rhyme scheme and write your own limerick!

limerick rhyme scheme	limerick example
a →	Mandy's cat, a tabby named Bag-o-tricks,
a →	Was fond of reciting old limericks.
b →	And then came a day
b →	When the cat ran away.
a →	All Mandy could say was, "Oh, fiddlesticks!"

The poet Edward Lear, pictured above, popularized the limerick in the 1830s with his *Book of Nonsense*. Limericks are thought to have originated in Ireland and are named for the city of Limerick, Ireland.

What Is Poetic Meter?

Poetic meter is the **rhythm** created by a pattern of strong and weak, or stressed and unstressed, syllables. This rhythmic pattern creates a musical quality, almost like the beat of a drum. Many poetic forms are based on predetermined metrical patterns.

English words are formed by patterns of stressed and unstressed sounds. Every word with more than 1 syllable has 1 syllable that is stressed more than the other syllables. Stressed syllables sound slightly longer and louder than the other syllables. Take the term "poetic meter," for example. The word "poetic" has 3 syllables, and the second syllable is the one that is stressed. The word "meter" has 2 syllables, with the stress on the first syllable. Say the words aloud. Can you hear the stressed syllables?

In poetry, a special system shows which syllables are stressed and which are unstressed. Stressed syllables are usually marked with this symbol: /. Unstressed syllables are usually marked with this symbol: ⌣. Using this system, the term "poetic meter" would be marked with this pattern:

⌣ / ⌣ / ⌣
po e tic me ter

Poetic meter can be used to set the pace in a poem. Many stressed syllables in a row can slow down the pace and make the poem sound "heavy" or "slow," as in the phrase "big black rock." A pattern that includes several unstressed syllables in a row can make the pace of the poem "light" and "quick," as in the phrase "merrily hurrying."

symbol for stressed syllable	symbol for unstressed syllable
/	◡

Peter Piper picked a peck of pickled peppers.

A peck of pickled peppers Peter Piper picked.

If Peter Piper picked a peck of pickled peppers,

How many pecks of pickled peppers did Peter Piper pick?

A tongue twister is a phrase, sentence, or poem that is difficult to say because of a series of repeated consonant sounds. Read the tongue twister above and listen for the stressed and unstressed syllables. The pattern of stresses creates a rhythm that is almost like a song.

Chart 1

name of poetic foot	symbols	example
iamb (I-am)	◡ /	because (bih-KUHZ)
trochee (TROH-kee)	/ ◡	able (AY-buhl)
anapest (AA-nuh-pest)	◡ ◡ /	disagree (dis-uh-GREE)
dactyl (DAK-tuhl)	/ ◡ ◡	terrible (TAIR-uh-buhl)

Chart 2

name of line	number of poetic feet
monometer (muh-NAH-muh-tuhr)	1
dimeter (DIH-muh-tuhr)	2
trimeter (TRIH-muh-tuhr)	3
tetrameter (teh-TRA-muh-tuhr)	4
pentameter (pen-TA-muh-tuhr)	5
hexameter (hek-SA-muh-tuhr)	6

The basic unit used to measure poetic meter is called a "foot." A poetic foot consists of two or more stressed or unstressed syllables. Each combination has a specific name, which depends on its pattern of stressed and unstressed syllables. Chart 1 on page 16 shows some common poetic feet and their names.

Poetic feet are linked together to form lines. A line of poetry can be measured by the number of feet it has, or by the number of stressed syllables it has. Chart 2 on page 16 shows the names of the different kinds of lines and how many poetic feet are in each.

Poetic feet and lines of meter were invented and named by Greek poets many centuries ago. "Meter" comes from the Greek word for "measure." Poets combine feet and lines to create poetic forms. A common form of meter in the English language is iambic pentameter. "Penta" means "five." The lines of a poem written in iambic pentameter have 5 iambs, or 5 feet of 2 syllables (unstressed-stressed), for a total of 10 syllables per line. Many English writers believe that this meter most resembles the sound of natural speech. The sentence below is written in iambic pentameter:

˘ / ˘ / ˘ / ˘ / ˘ /

He came to fix the table right away.

Do you hear the alternating stressed and unstressed syllables in this sentence?

Poetry written in iambic pentameter does not always have exactly 10 syllables and 5 stresses in each line. Some lines may have 11 or 12 syllables, or 4 stressed syllables instead of 5, or may include poetic feet other than iambs. Most poets would agree that a poem written completely in 1 form of meter sounds too **monotonous**. In other words, the pattern of beats created by the words is too repetitive.

Read the following lines aloud. Notice which syllables are stressed and which are not.

Friends, Romans, countrymen, lend me your ears!
I come to bury Caesar, not to praise him.

The first line tells us that the speaker is talking to a large group of people. Notice that the line, written in iambic pentameter, is irregular, with more stressed syllables than unstressed syllables. This emphasizes the idea that the speaker is shouting so that everyone he is addressing may hear him. Now count the syllables in the second line. While the second line is also written in iambic pentameter, it has an extra syllable at the end. The author varied the length of the lines this way to help create natural-sounding speech.

Can you hear the stressed syllables in the 2 lines above? Here is what the lines look like when marked with the symbols for stressed syllables and unstressed syllables.

/	/	˘	/	˘	˘	/	/	˘	/	
Friends,	**Ro**	**mans,**	**coun**	**try**	**men,**	**lend**	**me**	**your**	**ears!**	
˘	/	˘	/	˘	/	˘	/	˘	/	˘
I	**come**	**to**	**bu**	**ry**	**Cae**	**sar,**	**not**	**to**	**praise**	**him.**

Mark Antony, shown in the above painting holding the cloth that covered Julius Caesar's body, is the speaker of the lines on the opposite page. This powerful Roman general lived from about 83 B.C. to 30 B.C.

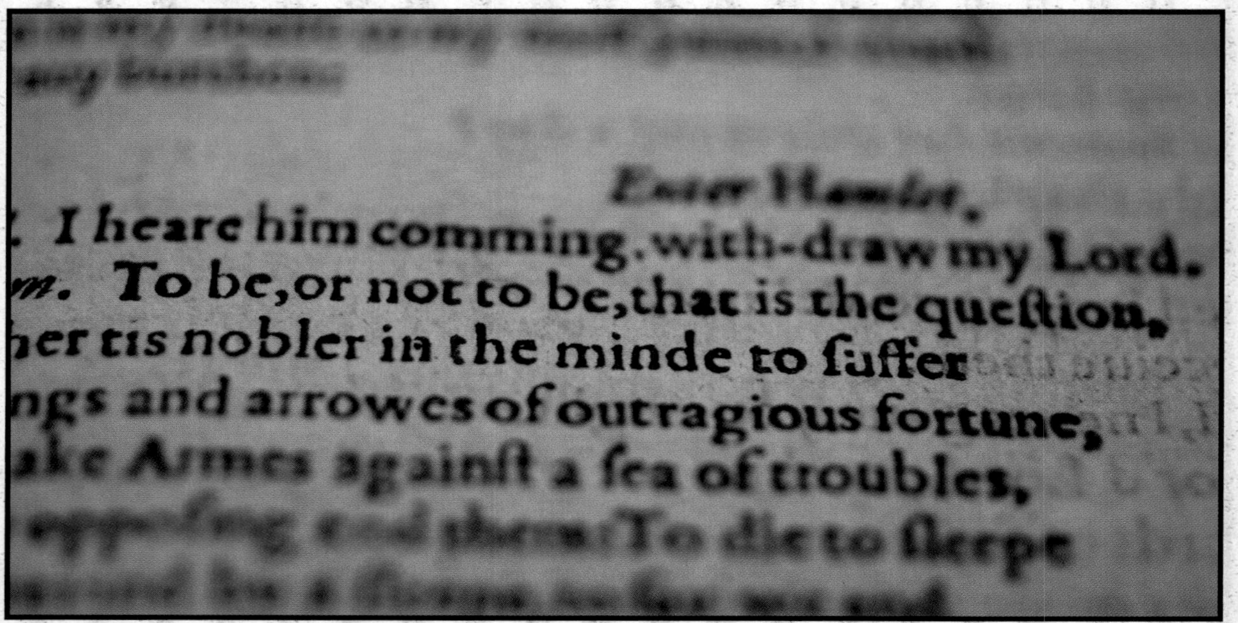

A rare copy of William Shakespeare's *Hamlet,* shown above, was displayed at Christie's auction house in London on March 1, 2004. Although this book printed in 1611 is only part of a large collection of early English drama, it is expected to be auctioned for between $1.5 million and $2 million.

The Poetry of William Shakespeare

The lines of poetry in the previous chapter, which come from a play titled *Julius Caesar,* were written over 400 years ago by one of the most famous English authors of all time, William Shakespeare. Shakespeare was born in Stratford-upon-Avon, England, in 1564. As a young man, Shakespeare moved to London. He became a famous actor, even performing for Queen Elizabeth I. Soon, he began to write plays for his theater company. Today, he is still one of the most admired authors in the world, almost 400 years after his death.

Shakespeare is often called the Bard, a term that describes a poet who writes about heroes and legends. His dramas and characters are well known for reflecting the true nature of humankind. Shakespeare's **comedies** focus on humor and love, his **tragedies** show the sorrowful side of life, and his historical plays feature famous leaders of the past. He is also well known for writing a type of poem called a **sonnet**. Although he wrote most of his plays and poems in iambic pentameter, Shakespeare used a variety of forms and meters to emphasize his ideas.

Many scholars believe Shakespeare was born on April 26, 1564. He died exactly 52 years after this date.

Let's look at a line from a Shakespeare play that is written in iambic pentameter. This example comes from a historical play titled *Richard III*. Read the line aloud and listen for the stressed syllables.

A horse! A horse! My kingdom for a horse!

The scene in which this line appears takes place on a battlefield in England. The speaker of this line—King Richard III—is losing the battle. His horse has been killed, and he searches for another horse to continue the fight. The repetition of the iambic phrase "a horse" helps to emphasize the **urgency** and desperation of Richard's situation.

When you read the line aloud, did you hear the stressed syllables? Notice that the line has only 4 stressed syllables instead of 5, even though it is written in iambic pentameter. However, the pattern of stressed and unstressed syllables still sounds like iambic pentameter. The fact that the line is not a perfect iambic pattern makes it sound more natural, since the word "for" would not be stressed in everyday speech. The irregular pattern of iambs also reflects the turmoil of the battle scene. Look at this line again with the syllables marked.

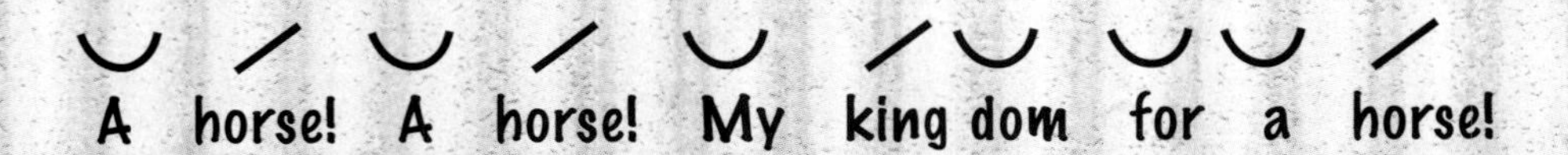

J. Chapman
sculpsit
RICHARD III
Publish'd as the Act directs July 1st 1799.

Shakespeare's play *Richard III* tells about the War of the Roses (1455–1485), a civil war in England between the families known as the House of York and the House of Lancaster. Both used a rose as their family symbol. In 1485, Richard III died on the battlefield, bringing the War of the Roses to an end.

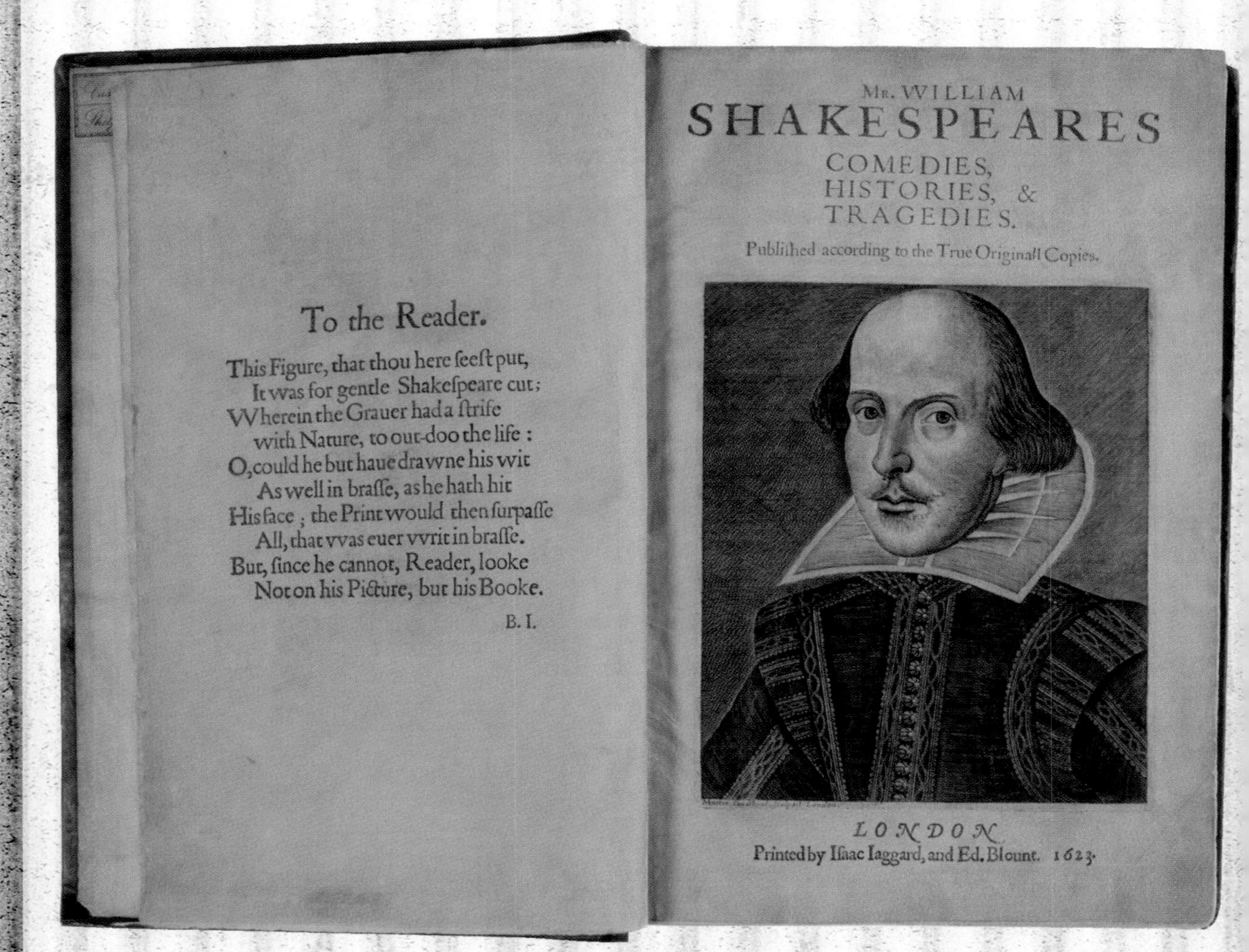

To the Reader.

This Figure, that thou here ſeeſt put,
It was for gentle Shakeſpeare cut;
Wherein the Grauer had a ſtrife
with Nature, to out-doo the life:
O, could he but haue drawne his wit
As well in braſſe, as he hath hit
His face; the Print would then ſurpaſſe
All, that vvas euer vvrit in braſſe.
But, ſince he cannot, Reader, looke
Not on his Picture, but his Booke.

B. I.

Mr. WILLIAM
SHAKESPEARES
COMEDIES,
HISTORIES, &
TRAGEDIES.
Publiſhed according to the True Originall Copies.

LONDON
Printed by Iſaac Iaggard, and Ed. Blount. 1623.

Shakespeare wrote 154 sonnets. His poems explore several common themes, including love, beauty, poetry, and the effect that the passing of time has on these things.

In addition to his many plays, Shakespeare popularized the English sonnet, which is often called the Shakespearean sonnet. The English sonnet is a 14-line poem written in iambic pentameter. It follows this rhyme scheme:

a b a b c d c d e f e f g g

As discussed before, this letter pattern means that lines marked with the same letters end with rhyming words.

Some of Shakespeare's sonnets are as well known as his plays. These lines are from "Sonnet 18," one of the most famous of all of Shakespeare's poems.

Shall I compare thee to a summer's day?
Thou art more lovely and more temperate.

The speaker in this poem is saying that the person to whom he is talking is more beautiful than a summer day. The speaker goes on to say that while summer's beauty eventually fades into autumn, the person to whom he is speaking will never lose beauty because the poet has captured it in the lines of this poem. As long as there are people to read the poem, the person to whom he speaks will always remain beautiful and young. The rhyme scheme and the pattern of iambs in this poem create a musical quality when the lines are read aloud. Can you hear it in the lines above?

Shakespeare did not always use iambic pentameter in his poetry. Like all great poets, he used different metrical patterns for different themes and emotions. The lines below are from one of Shakespeare's most famous tragedies, *Macbeth*. This play is a story about kings and queens, ghosts and witches, love and murder. These lines are spoken by 3 witches stirring a mysterious mixture. Read these 2 lines aloud and listen to how they sound.

Double, double, toil and trouble;
Fire burn, and cauldron bubble.

The lines begin with stressed syllables, which grabs the listener's attention. They are written with trochees (stressed-unstressed), not iambs (unstressed-stressed). If iambs are thought to reflect natural speech, then trochees might be thought of as "irregular" speech. What better way to reflect the speech of witches than to use an irregular pattern of speech! The lines each have 8 syllables coupled into 4 feet. The pattern of these lines is called trochaic tetrameter (troh-KAY-ihk teh-TRA-muh-tuhr). Look at these lines with marked syllables.

/ ˘ / ˘ / ˘ / ˘
Double, double, toil and trouble;
/ ˘ / ˘ / ˘ / ˘
Fire burn, and cauldron bubble.

This painting titled *Macbeth, Banquo, and the Witches on the Heath* was created about 1794. Macbeth grips his sword in fear as he and his friend Banquo encounter 3 witches. They tell Macbeth that he will become king. Hearing this, Macbeth kills the existing king and many others who stand in his way to power.

Homer, pictured on the opposite page, was an eighth century B.C. Greek poet. He is said to have written *The Iliad* and *The Odyssey*, which are long poems about ancient heroes. In the painting above, inspired by *The Iliad*, the citizens of Troy move the famous "Trojan horse" into their city. Greek warriors, who will attack and capture the city, are hidden in the horse.

Why Are Patterns Important to Poetry?

The earliest forms of poetry were songs memorized and sung by people thousands of years ago. This was a way for people to pass legends and history from one generation to the next. Today, poetic techniques like form and meter help to create a musical effect when poetry is read aloud.

Poetry that does not follow patterns of form and meter is called "free verse." There are no rules for writing free-verse poetry; the lines do not have to rhyme or be the same length and the author does not have to use meter. Free-verse poetry often contains elements like rhyme and repetition, but an author does not have a set pattern to follow when writing it. Just as the term suggests, the author is free to make up the form of the poem as it is written, instead of following a predetermined pattern.

Even free verse, however, can benefit from form and meter. As we have seen, language and the way it sounds affect the way our minds form thoughts and emotions. This happens all the time, but we usually do not think about it. Poetry, however, compels us to really listen to language and the patterns of sounds that it makes. Good poets use these patterns of sounds to enable us to better understand the ideas they want to convey.

Homer

All good poets know how to use language to create emotions and themes. Form and meter are just 2 tools to accomplish this. Poets also use methods like grammar, imagery, **metaphor**, and **simile** to express themselves.

If you want to be a poet, it is important to know that you don't have to use any or all of these methods when writing poetry. If you're not sure how to use form and meter to say what you want to say, just write and see what happens. As you gain confidence in your writing, however, you may want to practice with form and meter. It can be a challenge to use the patterns discussed in this book and the many others that exist. You may try to invent your own poetic patterns. The only limits are those that you decide to make for yourself. The only rules when writing poetry are to write and to have fun doing it!

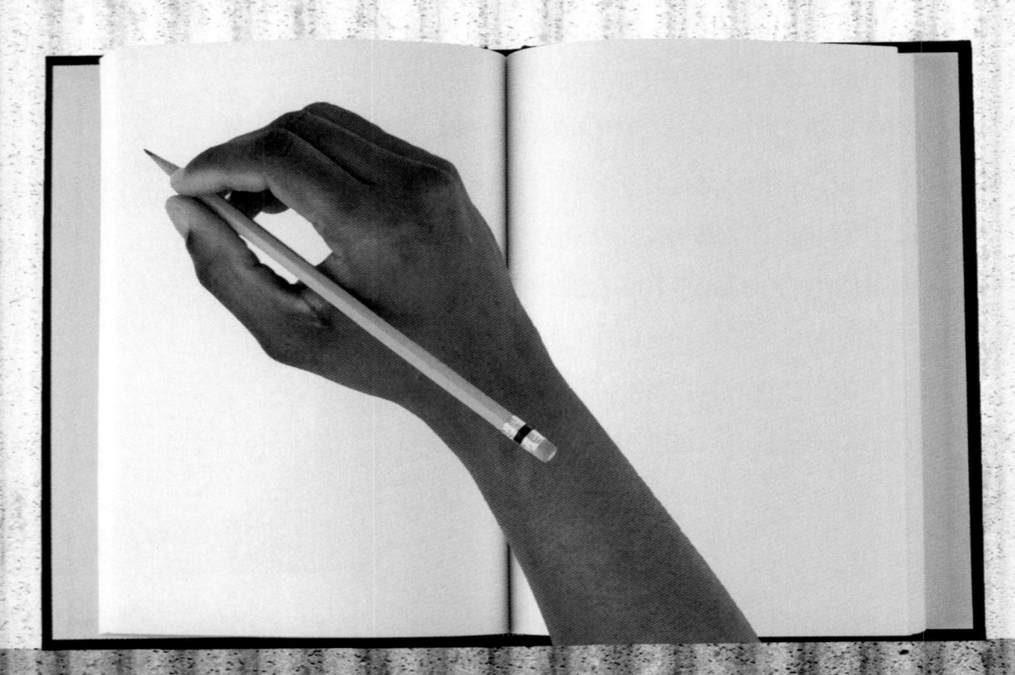

Glossary

comedy (KAH-muh-dee) A play that is light and funny, with a happy ending.

metaphor (MEH-tah-for) A comparison that does not use "like" or "as."

monotonous (muh-NAH-tuh-nuhs) Unchanging and unexciting.

predetermined (pree-dih-TUHR-muhnd) Already decided.

predictable (prih-DIK-tuh-buhl) Likely or expected based on what has already been observed.

rhyme (RYM) Two or more words that end with the same sound.

rhyme scheme (RYM SKEEM) The pattern of rhyming lines in a poem.

rhythm (RIH-thuhm) A repeated pattern of strong and weak beats.

simile (SIH-muh-lee) A comparison that uses "like" or "as."

sonnet (SAH-nuht) A type of poem that has 14 lines and follows a rhyme scheme.

syllable (SIH-luh-buhl) A unit of spoken language. It can be used alone or with other syllables to form words.

tension (TEN-shun) A state of emotional unrest or conflict.

tragedy (TRA-juh-dee) A play that is sad and serious, with an unhappy ending.

urgency (UHR-juhn-see) A state calling for an immediate response or action.

Index